Magic, Science, and Civilization

Number 20
Bampton Lectures in America
Delivered at Columbia University

Magic, Science, and Civilization

J. Bronowski

1978
Columbia University Press
New York

Library of Congress Cataloging in Publication Data

Bronowski, Jacob, 1908–1974.
  Magic, science, and civilization.

  (Bampton lectures in America; no. 20)
  Includes bibliographical references.
  1. Science and civilization—Addresses, essays,
lectures. 2. Science—Europe—History—Addresses,
essays, lectures. 3. Philosophy—Europe—History—
Addresses, essays, lectures. 4. Magic—Europe—
Addresses, essays, lectures. I. Title. II. Series.
CB151.B89 1978   909.08   78-1660
ISBN 0–231–04484–4

Columbia University Press
New York   Guildford, Surrey

Contents

Acknowledgments

Grateful acknowledgment is made to the following individuals and publishers:

To Harcourt Brace Jovanovich, Inc., and Hart-Davis MacGibbon/Granada Publishing Limited for permission to reprint "pity this busy monster, manunkind," by e. e. cummings, copyright 1944 by E. E. Cummings, copyright renewed 1972 by Nancy T. Andrews. From *Complete Poems 1913–1962*.

To Macmillan Publishing Co., Inc., M. B. Yeats, Miss Anne Yeats, and the Macmillan Co. of London and Basingstoke for permission to reprint "Fragments," from *The Tower*, by W. B. Yeats. From *The Collected Poems of W. B. Yeats*, copyright 1933 by Macmillan Publishing Co., Inc., renewed 1961 by Bertha Georgie Yeats.

To W. W. Norton and Company, Inc., and The Hogarth Press, Ltd., for permission to reprint lines from *Duino Elegies*, by Rainer Maria Rilke, translated by J. B. Leishman and Stephen Spender, copyright 1939 by W. W. Norton and Company, Inc., copyright renewed 1967 by Stephen Spender and J. B. Leishman.

To Suhrkamp Verlag for permission to reprint lines from "Furcht und Elend des Dritten Reichs," by Bertolt Brecht.

I
Interpretations
of Nature

IN SPITE OF my rather forbidding title, I propose to discuss what seem to me to be contemporary issues—namely, the place of science in the total field of knowledge and that of human values as we have to reformulate them in this century. I call that a contemporary issue; indeed I think it's the prime contemporary issue. At the same time I should say that I regard as from now on timeless the problem of what place science has in human thought and in human action. Now it may seem strange to say at the same time that something is a very live, present, contemporary, new issue and yet to claim that it is universal and timeless. But it does seem to me most important to grasp that both those statements are true. Something has happened regarding the eruption of science into Western Civilization which is new and yet permanent and irreversible. In the evolution of any animal there come moments when the species takes a radically new step, a mutation is built into the total genetic complex, and from that moment on the species is committed to some new way of life—like coming out of the water onto the land. Now I believe that the scientific revolution has done exactly that kind of thing to our cultural history and we must simply face this fact. It isn't just that science has happened, that if you don't look it might go away. There has been an irreversible step in the cultural evolution of man; it took place at the beginning of the scientific revolution from, say, 1500 to 1700, and it will never be undone. We are committed to a scientific way of thinking and to what it entails, a technological way of acting, and we cannot go back. I think that needs saying in this kind of firm way because particularly during our own lifetime scientific change has been rapid and headlong, and yet has seemed to creep up by small steps. And it is only when we look back to the great change between 1500 and 1700 that we become aware that we shall never go back to the pre-1500 days. Science may be destroyed, our civilization may change its character, but you cannot simply put that evolutionary clock into reverse. The step that was taken in the scientific revolution, which is a shorthand that I use for a whole complex of changes between 1500 and 1700, was just as radical as the invention of agriculture, the invention of writing, the invention of poetry and art, or the invention of urban life. All of those are things which we now take for granted in our civilization, but all were irreversible steps, and from the moment that they took place human life changed and nothing could turn it back. No amount of talk about country ways and how nice spring is in the woods will change the pattern of urban living. No amount of nature talk will change us from people who read and write and among other things read and write poetry and make works of art.

Forgive me for laboring this point, but so much has been said about the importance of science and so little has been said about this radical threshold that we have crossed that it seems to me important to say it. Indeed, I stole some years ago a phrase about this which I am now happy to acknowledge, because when I stole it nobody else quoted it and so it seemed to me perfectly all right to appropriate it. The phrase is that "the most remarkable discovery made by scientists is science itself" and my friend Gerard Piel, who in fact dropped this unconsidered trifle in conversation, has never reclaimed it. But now that I find that it is being quoted by other people, perhaps justice demands that I acknowledge its origin.

Now if that is true, then it is clear that like urban life, like writing and poetry, like agriculture, what has come upon us is something which wholly transforms our outlook on the rest of the world and in particular the relation between man and nature. That is essentially the theme that I am going to be dealing with in this volume and why I call my first chapter "Interpretations of Nature." Because science represents in some way a very general old human interpretation and in another way a special interpretation, that is the distinction that I must make.

But if indeed the difference is as radical as I have said, then no little nonsense about the second law of thermodynamics is really going to heal the kind of gap that has been portrayed. Fond as I am of Charles Snow, I must say that it seems to me slightly pathetic to go around with a little trowel and a little plaster pretending that you can mend the cracks between an old way of thinking and a new way of thinking by teaching people the second law of thermodynamics or, if you read the second edition,* give up the second law of thermodynamics and now concentrate on molecular biology. It isn't as simple as that—not only for nonscientists, but for scientists as well. There is a very considerable danger that scientists themselves, who become day-by-day more numerous, will give up the task of trying to make a whole human outlook and simply be content to retreat into their specialities. So that, much as I am naturally with Charles Snow as a professional scientist in throwing the blame on all those literary boys who won't read about molecular biology, it must be confessed that scientists bear a heavy and, in my view, an increasing responsibility from now on, for exhibiting the human implications not merely of what they do, but of their way of thinking. And if you want to think of an alternative title you could substitute "the scientific way of thinking is a human way of thinking"; and it is becoming for us the only human way which we can treat as a unifying discipline.

If science has made so radical a break during the last three hundred odd years, if it is now necessary to establish an outlook on man and nature which is founded in this scientific outlook, two plain questions arise. The first is, are revolutions really as intellectual as this? And secondly, are we really unique in trying to build a unitary system of the world?

*C. P. Snow, *The Two Cultures: A Second Look* (New York: Cambridge University Press, 1969).

5   Interpretations of Nature

Let me take the first question first. T. S. Eliot in writing the introduction to an edition of Seneca has a delightful phrase: "Few things that can happen to a nation are more important than the invention of a new form of [you'll never guess], . . . verse."

It's a startling remark and yet it's not wholly without justice. It seems to me to fit appropriately into this context because we really ought to face the fact that in the end the great human revolutions are intellectual revolutions. That what happened, to take up Mr. Eliot, to the heroic couplet in England in the late seventeenth and early eighteenth century represents a new approach, not merely to verse, but to argument, to scientific and non-scientific thinking, and it is important. However, recondite literary criticism aside, if Mr. Eliot has a right to say that about a new form of verse then we certainly have a right to say it about that new form of thinking which came in with the couplet at the time of the Royal Society in the last part of the seventeenth century. So I am wholly unashamed to be discussing intellectual issues. I shall discuss what is essentially an intellectual revolution and I am unashamedly an intellectual. Indeed, I know no other way to be a human being.

Now I turn to the second question. We are faced with having to try to build a unitary theory of the world—one which involves all human life and yet is conceived in the scientific discipline which, in my view, has transformed the world since 1600, plus or minus a hundred years. In my view, this is a unique endeavor. There has never before been a moment in history when there has been an attempt to see the world as whole and from a single core or type of explanation.

Now obviously as a historical remark that is bound to be open to challenge, and the most obvious challenge is to say, "oh come now, from 1250 until 1550 there was dominant in central Europe an outlook based upon the writings of Thomas Aquinas in which were unified the doctrines of the church and Aristotelian physics and philosophy. Is it not true that that aimed for and in some sense achieved a single unitary outlook on the world?" I don't think that's true. And I think the best way to show that it's not is to look at a specific problem that we all know and see what happened to it during that time—the law of gravitation.

When Newton published the *Principia* in 1687, it was a sensation—not only among scientists, but also among all thinking people, who instantly recognized that the world had been reorganized. Alexander Pope, using the heroic couplet I have mentioned, said, "nature and nature's laws lay hid in night,/ God said let Newton be and all was light." And that's really what people felt; Bentley, the Master of Trinity, the greatest panjandrum in the church of England at that time, said that was why he had asked Newton's permission to preach on the law of gravitation as the last and ultimate example of the divine ordinance.

However, *Principia* almost was not published. Since it was based in large part on Newton's statement that gravitation obeyed a law of inverse squares, there was a quarrel between Isaac Newton and Robert Hooke. Hooke said that he would like some acknowledgement in the Preface that he had also thought of the law of inverse squares. Newton, with that peremptory grandeur which he always had when he was in an unassailable position, said, "insist and I shall not publish." So there is no acknowledgement to Hooke.

We are now talking about the 1680s and it seems strange that we should be arguing about whether Mr. Robert Hooke or Sir Isaac Newton was really the person among a host of others who proposed that the gravitation between two massive bodies falls off as the inverse square of the distance. Because if you think back for a moment, to that famous scene in 1666 in the country garden when Newton was a boy, the fact that the law was an inverse square was almost incidental. What Newton really did in the country garden was to calculate what would happen if gravitation reached to the moon. In other words, the problem in 1666 was not the form of the law of gravitation but the mere notion that it reached as far as the moon. Then came Newton's much bolder leap that it reached the sun and beyond.

Now that was a fairly new speculation, but not entirely new. Because in 1618–19, give or take a few years, Kepler had published a very extraordinary book called the *Somnium*, which was an imaginary journey to the moon—a great favorite of science fiction throughout the seventeenth century. But Kepler was the first man who said that when you travel from here to the moon, gravity will go with you all the way, but there will come a moment when the earth's gravity will leave off and the moon's gravity will become the more important. That's a gorgeous thing to have said 350 years ago—especially when you consider that it took almost exactly 350 years for men to experience a phenomenon which this funny little astrologer, who made a bare living writing books about astronomy, had already talked about in 1618 or thereabouts.

If you read the *Somnium*, Kepler says there is a point when gravity will fall off, and then he does a calculation and the calculation is wrong. The calculation is wrong because he believed that gravity fell off as the inverse distance, not as the inverse square of the distance. Now there are very good reasons why he should have made this mistake. If one really thought seriously, there were only two good guesses—the inverse distance or the inverse square—and one in two is about as well as you can expect to do.

It's a puzzle to ask how did Kepler come upon this notion. The book did not appear until after his death, when it was published by his son. It got his mother into trouble about witchcraft, since the local clergy believed that it described an actual journey to the moon—which, of course, 350 years ago was treated as disgraceful and not, as today, as heroic.

It's not very clear where Kepler hit upon this notion, because it was well ahead of its time—roughly fifty years ahead of Newton and his time. But it seems likely that he had been reading some neo-Platonists, who were fashionable then. Among those neo-Platonists he read Nicolas of Cusa, who said something about the fact that massive bodies ought to attract one another. Now Nicolas of Cusa took this from a fifth-century writer called Dionysius the

Areopagite and Dionysius said that massive bodies are sure to attract one another because the whole universe is filled with the love of God: therefore every massive body is filled with love for every other massive body, and therefore all massive bodies are sure to attract one another.

If I had absolutely to choose between the *Principia* and Dionysius the Areopagite I am afraid I should be sorely tempted, at any rate for some minutes, to say that we ought not to lose that marvelous speculation which Pierre Duhem dug up. However, I have quoted this special example in order to make two points about it. One is that you can't do any science with the love of God. Dionysius had an excellent idea—bodies attract one another—and he had a good reason to think so. But it was not a reason which was going to tell him whether the attraction fell off at the inverse distance or the inverse square. It was not a reason with which you could do any kind of manipulation. It was not a coherent reason, because it lent itself also to a different kind of speculation. The following quote is from a follower of Pico della Mirandola and Savonarola. This is the great neo-Platonic tradition. It is from the year 1487, found in a book by Nesi on *De Charitate Dei*, of the Love of God.

> I transform the lover into the beloved and the beloved into the lover. The lover becomes the beloved because the lover dying, lives in the beloved. And the beloved becomes the lover for he learns to know himself in the lover and gets to love himself through the lover. And while he is thus loving himself in loving the lover he loves the lover who himself has become the beloved.

Now I am not citing this passage entirely out of mischief—although I readily grant that an element of mischief has gone into my seeking so extravagant a quotation—but in order to make it clear that much as Aquinas and the Roman Catholic Church might struggle with this problem of founding physics on the love of God, or on other imagery which comes from a different field, the problem is insoluble. There are essentially two viewpoints. I think this is a splendid passage; it's just as good as anything in French existentialist literature. But like contemporary French existentialism, for which I have the greatest respect, it really won't solve any differential equations.

There is a second reason why the story about Dionysius the Areopagite makes it so clear that you couldn't possibly get a unified view of the world from that outlook, and it's a very simple one. When Kepler read Nicholas of Cusa, when Nicholas of Cusa read Dionysius the Areopagite, they all believed that there really was such a person. Well, there wasn't. The man who wrote that passage about the love of God was a fifth-century imposter who pretended to have lived much earlier than he did, so that his writings might be accepted in the church as being part of the law of the church fathers. This is a crucial issue to which I shall be returning in the last chapter, but I might as well make clear now that the problem with society before Copernicus, Kepler, and so on was that what you said didn't have to be true so long as it was religious. Now that seems a very harsh thing to say and I am not anxious to outrage anyone's religious sensibilities, but there was at

that time a way of thinking which the practitioners themselves called the way of the two truths—there was a truth
by faith and a truth by logic. It is of course clear that
Aquinas was trying to combine these, but from the time of
Aquinas, from 1250, to around 1500–1550, there were a
lot of documents which had been written into church
history that were simply fakes, and yet the people who
wrote them in did not think that this in any way outraged
the dignity of the church. There was an intellectual order
in which you could claim that you were Dionysius and
that you were virtually a contemporary of Christ although
in fact you lived five centuries afterward. And it's that
dichotomy which, in my view, made it impossible to
conceive a unitary view of the world at that time.

I have made this rather long digression because I think
gravitation and the love of God are all simply marvelous
subjects, but also because I want to make it clear that if in
fact it had not been possible even in the work of Aquinas
to establish a workable attempt at a unitary view of the
world, then I am justified in claiming that this is what we
are doing now, and that it is a unique attempt and a very
difficult one. It is very much easier to divide your outlook
on the world into two halves, to say that you know this
belongs to the daily half and this belongs to the Sunday
half. This belongs to how I have to work, but this is the
series of faiths by which I live. I call this divided view a
magical view of nature and man, and that's why I have
called these essays *Magic, Science and Civilization.*

My definition of magic is very simple. It is the view that
there is a logic of everyday life, but there is also a logic of

another world. And that other logic works in a different way and if you can only find the secret key, if you can enter into some magical practice—particularly if you can find the right form of words—then either the Almighty will be on your side, or you will collect all the votes, or people will believe that because you call it peace, that it's not the same word as war, and all those other things which Orwell has portrayed so brilliantly but which really always come to the same thing: trying to command the world and particularly the opinions of other people by some formula which is other than the truth.

In these essays I have given a leading place to the concept of magic, because we really are at quite a crucial moment. A new magic has been growing out of science itself. It's the magic of technology, it's the magic of really flying to the moon. After all, as Professor Franklin Ford of Harvard has put it so elegantly, watching people in orbit is now the most popular outdoor sport in the United States. And what is splendid about that is not watching them in orbit but the notion that they are carrying out some kind of activity which is romantically different from ours. Now this is the magic formula; but only the esoterics know it. Only the initiates have the key to it. And the key in general that is being offered us all now is some entry into a special form of technology. Magic is a technology, a technology without science; and technology does duty for magic. And our heroes now tend very much to be the magic heroes. I have seen this in my own lifetime—the spy, the agent provocateur, the woman who secretly passes a note across a railway counter on the borders of

13 Interpretations of Nature

Czechoslovakia and Russia, the double agent, the man who is shot climbing the Berlin wall. All these people carry this magic formula of God in their mouths, the way that the Prague legends say that Golem carried the four letters of God's name in his mouth, which is what made him work. And with this magic, boundaries will somehow be overleapt, science will be at a standstill, reason will be ended, and you will enter with a kind of halo into a universe of occult power. The <u>central opposition between magic and science is the opposition between power and knowledge</u>.

One quotation about technology will underline my point. The quotation comes from e. e. cummings.

> pity this busy monster,manunkind,
>
> not. Progress is a comfortable disease:
> your victim(death and life safely beyond)
>
> plays with the bigness of his littleness
> —electrons deify one razorblade
> into a mountainrange;lenses extend
>
> unwish through curving wherewhen till unwish
> returns on its unself.
>
>   A world of made
> is not a world of born—pity poor flesh
>
> and trees,poor stars and stones,but never this
> fine specimen of hypermagical
>
> ultraomnipotence. We doctors know
>
> a hopeless case if—listen:there's a hell
> of a good universe next door;let's go

I remember hearing e. e. cummings read some of his poems in the last lectures he gave at Harvard and being deeply shocked at the anti-science which came out of every word he uttered. And being shocked even at the time by the fact that he was so well informed. For example, "lenses extend unwish through curving wherewhen." That means through curving space-time made into a single four-dimensional unity, "through curving wherewhen until unwish returns on its unself." Now you remember that's the bit about how if you travel in a curved space-time, you come back on yourself. It's all terribly well informed; I have no doubt it only comes out of popular articles on science but it's very well done. If he has been reading the popular articles, by Jove, he has picked the good ones.

But it's wholly anti-scientific in its outlook, because all the things it picks out, "electrons deify one razorblade into a mountainrange," are all to do with the technological marvels. And, "manunkind" was not invented by scientists. And as for "there's a hell of a good universe next door, let's go," there is no such universe next door, and there is nowhere for us to go if we decide to leave this one. Whatever may be about curving wherewhen, this is the curving wherewhen in which we are fixed and no attempt to undo science by saying let's go into the woods and play will do. Well, then, what is the crucial issue; how are we going to cope with this? Here we are at the end of this long panorama, Copernicus dethroning man from the center of the universe, Wöhler showing that organic molecules are no different from inorganic molecules, Darwin showing that man belongs to the tree of evolution of other animals, Freud, Einstein . . . you can make up the rest for yourself. Here we stand at this moment; how are we to heal the breach between man and nature which people like e. e. cummings are pointing to, which they feel? I have already said that it cannot be patched up by scientists learning some of the more amusing e. e. cummings poems or poets learning some of the more amusing things about how four-dimensional space goes.

At the end of those steps, from Copernicus through Wöhler and particularly through Darwin and Mendel, we have to face the fact that there is only one way to do this, and that is to ask ourselves questions about man in which we display his nature as ruthlessly, but as thoroughly, as we do in the examination of any other species. After all, what makes man unique among species? It is precisely that he is the only creature who sees the world both inside and outside, who is capable of looking into his own motives and at the same time looking at other people as if they were, not himself repeated, but some alien species.

The subject which I have been practicing is called by me "human specificity." And what makes the human species special is, in my view, the crucial link in building a view of the world in which nature and man are really joined into a unity. One thing that makes the human species special is exactly the ability to see ourselves both as inside and as outside. We have the ability to make a unitary view just as if we were turning the topology inside out. There are a great many species-specific gifts, and I shall be discussing them. But the one on which I want to center here is the fact that human beings guide their conduct by making plans.

I shall not talk about anything as highbrow as free will—whether the plans that you make are really yours to make, whether the behavior that you believe to be an act of choice is really behavior which you can choose. We must concentrate on something much simpler and more direct, which is that we are the species that guides its conduct by planned choices. It may be that when people all assemble on an evening it is as a result of elaborate molecular interaction, which we only call plans. That's really a matter of absolutely no interest; that's a question of metaphysics which has nothing to do with an empirical examination of how human beings behave, because the fact of the matter is that they are there on an evening because they planned to be there on that evening. And they made those plans with a perfectly conscious object in mind, with a teleology which was really directed to that time, that evening, that place.

I shall not discuss any metaphysical questions about whether that mechanism is indeed machine-like or not, because those seem to me questions which have nothing to do with how we actually behave. We have been too long preoccupied with questions about how the machinery actually works at a level which we are not able to analyze instead of asking ourselves what behavior distinguishes one set of organisms from another?

Rats are distinguished from cockchafers not only by their physiology but also by perfectly well understood behavior patterns. And human beings are distinguished from rats, not only by their physiology, but also by a form of behavior to which the plan is central.

17   Interpretations of Nature

The most important form of plan that we have been able to evolve in practice over the last 300 years is, of course, the scientific plan. We understand the machinery by which inanimate objects work much better than we understand anything else, and we are therefore inclined to think that all actions must have that form of machinery; but where do values, where do other aspects of contact come in? I will say first that this is not a correct description of science as carried out by human beings (see chapter 3), and second that when you look at how human beings do in fact plan, then long-term strategies such as ethics, values, and other forms of behavior become absolutely crucial in determining how they shape their lives (see chapter 4).

In summary, first science is not an independent, value-free dissociated activity which can be carried on apart from the rest of human life, because second, it is, on the contrary, the expression in a very precise form of the species-specific human behavior which centers on making plans. Third, there is no distinction between scientific strategies and human strategies in guiding our long-term attack on how to live and how to look at the world. Science is a world view based on the notion that we can plan by understanding. Fourth, science is distinguished from magical views by the fact that it refuses to acknowledge a division between two kinds of logic. There is only one logic; it works the same way in all forms of conduct and it is not carried out by any kind of formula but by an active view of how you apply the logic of long-term planning strategies to the conduct of the whole of your life.

Finally, and most crucially, science is distinguished from earlier forms of trying to achieve a unitary view of the world by the fact that there is only one form of truth in it. There is no distinction between man and nature, there is no distinction between the logic of magic and other logics, and there is no distinction between means and ends.

At the end of five or six centuries of searching for the truth by all kinds of dogma if there is a single thing that we can say that science has which has been the drive and direction in its success, it is that it has never said, "ah, well you know, that's a very good end and let's not be so very particular about whether we call it inverse squares or simply inverse, you know, it's all in a good cause." Now, there is one thing about science which I believe to be crucial to any resolution of the world today, and that is an elimination of any appeal to means and ends. If killing people is bad, it's bad; there are no good ends for killing people. If telling lies is bad, it's bad; and no Dionysius is serving any good end by pretending that he was a contemporary of Christ. So that the particular form of unification which I think to be crucial is that science is the one world view which believes that once you have formed your outlook on life, your world picture, then means and ends must be one. That's what you are going to stick to and there are no other ends to hold to.

II
Black Magic
and White Magic

I ENDED CHAPTER 1, on interpretations of nature, by giving a listing of five characteristics of science as I conceived that interpretation of nature. First, science cannot be conceived as an isolated or independent activity because, second it must be thought of as a world picture. This is how from now on we shall see the world until someone comes up with an even more sophisticated view. And third, that world picture has the characteristic of being active; it thinks of knowledge as being directed toward planned action, the human species-specific concept on which it rests is that human beings are planners, that they guide their conduct by looking ahead and making choices, and that science is an expression of that quite general human approach. For that reason it is an integrated approach. You cannot in science, and this was my fourth point, make any bifurcation and say this part of life is scientific and specialized and that is not; nature belongs here, man belongs there; or means belong here but ends stand above all that over there. And finally, my fifth point—above all you cannot make a distinction between power and knowledge, and that's the central distinction which magic makes. There is no way of doing science which is magical. It is that fifth point that I shall elaborate upon here.

The phrase _knowledge is power_ was uttered for the first time so far as we know in 1620. Until that year people did not think about power—either about personal power or about the power over nature—in those terms. As I have stated, I call everything magic which dualizes our view of the world. The Cartesian division between mind and body, I think, is a piece of old-fashioned magic which we had better forget. There are similar divisions which haunt us now—between technology and science or politics and conduct—which I shall return to in later chapters. Here I am going to stress something about the magical view which we have forgotten in the course of history but which threatens us in our own contemporary outlook.

The form of magic that I shall discuss is the notion that there is a way of having power over nature which simply depends on hitting the right key. If you say "open sesame" then nature will open for you; if you are an expert then nature will open for you; if you are a specialist of some kind or if you are remote, if you are esoteric, if you are an initiate there is some way of getting into nature which is not accessible to other people.

Now this was the dominant theme of all those centuries up to the fifteenth. And all primitive forms of magic— sympathetic magic, the kind of magic you read about in Levi Strauss for instance, magic that structuralists talk about—all come back to this notion: there is a way of having power which is esoteric and does not depend on generally accessible knowledge. Now I think that is fundamentally false and I also think, of course, that it is terribly dangerous, because it recurs in every generation. But let me say something about it in this highly specific context of magic up to the fifteenth century.

One of the things that must have struck you if you have read any book about magic is that there is a tendency for the rituals of magic to turn nature upside-down. For instance, if you have ever seen an illustration of a witch riding a broomstick, she does not ride the broomstick sitting forward, she rides the broomstick sitting backward. Now it may seem a childish thing for an eminent intellectual historian to be discussing which way witches sit on broomsticks. But the fact is that intellectual history is made of exactly such points. Why did people think that satanic rituals had to be set backwards? Why did people celebrate the black mass by going through the mass in reverse? Because the concept of that conquest of nature was that whatever the laws of nature were, the magic consisted of turning them back. What Joshua said was, "Sun stand thou still"; he didn't say anything about ellipses and inverse square laws. He said "let us stop the laws of nature and turn them back in their tracks." And really, one could say, if I may put this terribly crudely, that until the year 1500 any attempt to get power from nature had inherent in it the idea that you could only do this if you forced nature to provide it against her will. Nature had to be subjugated, and magic was a form of words, actions, and pictures which forced nature to do something which she wouldn't of herself do.

Let me note here that science does exactly the opposite. But it is important to realize that the subjugation of nature is the theme of all magical practice. We must get her to do something for us which she wouldn't do for everybody else—which means we must make her disobey her own laws. Of course, people before 1500 didn't really have much of an idea of what a law of nature was. But insofar as they conceived of nature following a natural course, magic was something which reversed it.

[What I am saying is my view, although it is not the view of all those who have written about magic. Lynn Thorndike, for example, was an eminent writer on the subject. His eight-volume work on the subject is entitled *The History of Magic and Experimental Science* (New York: Columbia University Press, 1923–58). It would be impertinent of me not to state that he thought differently. However, the very title *The History of Magic and Experimental Science* implies a view of science which is different from mine. What Lynn Thorndike said was that there are in magical and particularly alchemical practices many techniques which later formed an important part of technology and experimental science. Now that's undoubtedly true. But alas, in my view, this has nothing to do with the case. Of course there were people of all kinds practicing all kinds of alchemy right up to the days of Newton, whose alchemical writings are so voluminous that they were never published. Nevertheless, my main interest is in their attitude toward how the world works and how you make it obey, and not at all in their discoveries of how you smelt this or how you make that process in metallurgy work. It was the view of Lynn Thorndike, and it has been the view of some other eminent historians of science, that there is a continuity running from even before the Middle Ages into modern science. This is what Pierre Duhem was anxious to show, and in a way this is what George Sarton said also. And of course, there is some truth in that. There is not the slightest doubt that any particular piece of science that you have today can be traced to some fantasy in the Middle Ages. I did it in chapter 1, when I noted that the law of gravitation goes back to what the pseudo-Dionysius said about the love of God filling all matter.

But it is my view that those continuities give a false perspective of the great threshold from which the burst of modern science comes. And I would put this quite simply: I don't know whether science was born before 1500 or not (though I don't believe it was) but I do know that, mysteriously, magic in fact died after 1500.

I ought also just to pay a small obeisance to those historians who think that we ought not to look at the history of the Middle Ages or the Renaissance as if in some way it were the forerunner of today. I wrote a book of intellectual history and was amused to find that one of my kinder critics said that it was all very fine, but why did I think that the present age was any better than the fifteenth century? Well, I don't know whether it is better, but it seems to me terribly interesting that the fifteenth century has led to the present age and that the present age has not led to the fifteenth century.

My view of history is essentially an evolutionary one. I think it is right that we should look at history with hindsight, because I think, for two reasons, that the most important species-specific thing which man possesses and which started him off on his evolutionary career is exactly hindsight. If you make any plans, only hindsight will tell you whether they were any good. Secondly, we know from work on memory that it is only from hindsight, only from memory, that imagination and foresight develops. So I make no apologies for the fact that I shall discuss the history of the past as if the most exciting thing about it is that it has led us to the present.

I have said in chapter 1 that a point which must be made very forcibly about science is that it took an irreversible step in the cultural evolution of man. I have noted that we did not lay enough stress on the fact not only that science has made our lives different but also that it was a threshold of this kind. Holding that view I am bound to say that the dates of the scientific revolution between 1500 and 1700 do represent a major threshold in the development of science.

Now that seems strange to people who have tried to trace the history of science back beyond that time, because they point out that, after all, there was a school of people—who read Aristotle, who were Averroists, who continued to talk about scientific truth and distinguish it from spiritual truth—in a number of universities, such as Paris and Padua, through the thirteenth, fourteenth, and fifteenth centuries.

The school that looks for a continuity in the development of science looks for it there. Now I think that view is mistaken because I think that in about 1500 something very remarkable happened in all intellectual history, of which science is a part and a crucial part. It's not terribly fashionable to talk about the Renaissance now, because everyone is very busy explaining how it all really started much earlier. And it's not very fashionable even to talk about humanism, because very eminent scholars, including Professor Kristeller of Columbia University, have pointed out that humanism was a special kind of academic syllabus that led to the elaboration of rhetoric and theories of language at the expense of theology and other practices. In itself, humanism did not make a new way of life, and of course it appears to have had no influence on science. All that seems to me to be quite right. And yet it is absolutely true that Florence in 1500 was a different city (I cite Kristeller again) from Florence in 1400. Something had happened in Italy which made a great inroad in established, authoritarian, and traditional views of life.

When we come to revalue the Renaissance over the next twenty or thirty years of scholarship, the view that we are sure to come up with is that the most important thing was not that people in Florence started reading Plato instead of Aristotle, or that people from Padua argued about this or that, or that Ficino wrote this and Pomponazzi wrote that, but that in some way a dissolution of tradition took place and there developed an interest in new things in which the particular character of the new was not nearly so important as the shaking up of the old. And that characteristic was crucial to the development of science at that particular time. To my mind, the most extraordinary thing is that about 1500 the incursion of neo-Platonic and mystical ideas gave that impulse to the human mind, made that intellectual revaluation from which science and the arts took off together. The view that I am putting forward is that this revolution worked as much in the sciences as in the arts and that it is impossible to understand the really radical change that the Renaissance made unless we see science not as an afterthought but as an integral part of that humanism—rhetoric and linguistics and all.

Now to some small, interesting, specific examples: Between 1450 and 1465 Cosimo de'Medici began to collect a library of Greek manuscripts. They were being brought to the West by scholars and he sent his own merchants out to buy them up. They brought back the *Dialogues* of Plato, which had still not been translated from Greek, and they brought back also an incomplete manuscript of the *Corpus Hermeticum*, the fabulous book of magic of the Middle Ages of which, again, only a small part had been translated into Latin. His secretary was a man called Marsilio Ficino, and in 1463 he was translating the dialogues of Plato when Cosimo de'Medici told him to translate the *Corpus Hermeticum* first. In fact, Cosimo died the next year and he obviously felt that this esoteric knowledge, this magic, he had to know. Now the *Corpus Hermeticum* is an extraordinary book which remains in our language simply because we still use the phrase "hermetically sealed" to mean sealed by secret alchemical formulae. Although the book is called the *Corpus Hermeticum* because it was supposed to be a book about Hermes Trismegistus (Hermes is the Three Times Great who was supposed to have been a mythological Greek-Hebrew character), Ficino thought he might have been Moses himself. I need not tell you that the books are fakes, but that was not discovered until 150 years later—and anyway, as I have already explained, fakery was no crime at that time.

Ficino was an extraordinary character because he came at a moment when the old black magic, the witches' Sabbath, and so on, was not done anymore. Ficino never got up and said "I won't go to a coven of witches." He was too polite, too much of a gentleman, to do all this romping about in the nude in damp fields with satanic imagery and goats, and he was from a new kind of upper class society that was taking an interest in magic. They were sophisticated and gentlemanly, and this was not the kind of black magic in which they could be interested. And yet, Ficino really did sing hymns; he really did believe that he was conjuring down the influences of the planets and that in some way the world was opening up, that Orpheus and Pythagoras and all those planetary influences were one.

This is really the central point of neo-Platonism as Ficino introduced it. The world is a great harmony, and harmony is the crucial word. It literally meant both music and mathematics and incidentally also poetry. All these things were different aspects of the universal spirit, the *anima mundi*, the thing that Plato and Plotinus said was like a great organic creature of the world. Indeed Giorgio said that all this was simply a description, that the universe was the face of God and that all its aspects—music, poetry, and mathematics—were different expressions of the fact that it was a harmonious whole. Music and mathematics go together because Pythagoras and the Greeks, three thousand years ago, had discovered that in order to make an octave you have to make a musical string twice as long and in order to make the other main notes you have to have whole-number separations. And this extraordinary notion that the length of the vibrating string also gives pleasure to the ear and fills your soul with harmony had come down from the Greeks. For instance, Pythagoras invented the phrase "the music of the spheres." Shortly after the time I am describing, Kepler tried to fit the five Platonic solids into the orbits of the solar system because he naturally felt that all these things must go together—mathematics, music, harmony were one. Harmony is indeed the word to hold on to.

The exciting thing about these neo-Platonists is, first of all, that they made people interested in mathematics. It was from that moment onward that Greek mathematics was rediscovered, became exciting to people again; they started arguing about Playfair's axiom and all kinds of things that were unclear in Euclid. That led to natural knowledge through mathematics of which Newton really set the keystone. Secondly, people like Ficino had this marvelous sense that the world was at once both intelligible and beautiful. The phrase (which does not come from Ficino) about its being the face of God is the crux here. We have the sense that suddenly the Middle Ages were over. That rather heavy view of God sitting on the world with man quietly padding around making sure that he doesn't give offense had ended. There was suddenly a rainbow in the sky, the world was beautiful. We have the transcending sense of the beauty of nature, but above all of the beauty of the creation.

We see this outlook in Copernicus when in the next century, in 1543, he published the book that he had been working on for nearly thirty years about the revolution of the planets. He talked about the sun, about how marvelous it is. Of course, the textbooks just tell us that in effect he said, "well, it works simpler if we put the sun in the center of the universe." But that's not what he said. He said the sun was fit to be the center of the universe. The sun was marvelous. And he took that straight out of Ficino, who in fact wrote a book called *Of the Sun*. It has only recently been discovered that when Giordano Bruno came to Oxford in the next century and lectured about the Copernican system, the Oxford Dons treated him with grave suspicion, and particularly because they all spotted the quotations from Ficino that he thought they wouldn't know.

This sense that man and the universe are one, that the presence of God in the universe is a different kind of presence, is what makes the neo-Platonic revolution crucial in the science of the Renaissance. I have called it antiauthoritarian, but I ought really to have said antitraditional. Now I do not mean by this that people would suddenly go around saying God is dead, which would have been inconceivable then. What was happening was something quite different: there had been a hierarchy of God, man, nature. And that in that hierarchy God and man had moved into one position. Man was still dominating nature, but there was no longer the sense that he was under any higher authority. Everything that God had expressed was expressed in man.

We see this best in a follower of Ficino, Pico della Mirandola, who in 1487 proposed to dispute a famous series of theses, which have since come down to us under the title *Of the Dignity of Man*. Now, the dispute really was to a considerable extent about the dignity of man, provided that you understood this equivalence of man and God. Pico della Mirandola was saying above all that man was a unique animal because he was the only animal that made himself, that had no species-specific properties. Well, that is something of an exaggeration, but you know it is not quite as bad animal behavior as one would think, because it is certainly true that the most important part of the human equipment is its enormously greater flexibility and adaptibility than that of any other animal. In biology we generally express this by saying that whereas every other animal fits into an evolutionary niche, man essentially is busy hewing his evolutionary niche out of nature for himself.

That's not what e. e. cummings said in the poem that I quoted in chapter 1, but it is true just the same. It is just what makes me anxious to see that poets and nonscientists seize this—that the ability of man to make himself and particularly to make himself by thought is crucial. That comes out of Pico della Mirandola.

Pico della Mirandola was also very much against astrology. And he says so in the oration, which is full of all kinds of stuff out of the Cabala—all kinds of things that nobody reads anymore except in corners. And yet, he says in 1487, astrology is wrong. Why?—because it outrages the dignity of man that we should be subject to the influence of the planets out there traveling on their immutable, dreary predestined courses. That cannot be consonant with the dignity of man. It is a gorgeous thought; naturally we would put it rather differently now. We would say it's not consonant with the dignity of a *planet*.

Both Ficino and Pico as well as a number of other people about 1500 were practioners of magic, and yet their magic had a different quality. They were no longer trying to force nature into a different mode. In some way, they were trying to exploit a preordained harmony in nature. Ficino says this quite firmly: "When I sing a song to the sun it is not because I expect the sun to change its course but [because] I expect to put myself into a different cast of mind in relation to the sun." Now this is a very important concept that developed between 1500 and 1550—the notion that yes, there is a magic, but it is a *natural* magic, a *white* magic. No one knows quite how it works, but it attempts to extract out of the universe its own harmonies for our good. And here we are on the way to science as we understand it. If one had to put a date to this, one would say that roughly speaking between 1500 and the publication of Porta's book in 1558, which was called *Natural Magic*, the turning point took place. Of course, I am not trying to make the sort of point that on the 27th of February it all changed. History doesn't work like that. But what did happen was that highly intelligent people were bothered about demons and angels and all the oppositions in the old magic; they were convinced that the universe was harmonious, that man could be in contact with it, and they asked themselves how this could be done. There is a fascinating series of writings to be explored still between 1500 and 1550 (complicated of course by the occurence of the new humanism)—those of people like Erasmus, Luther (Protestantism) which keep on saying—"well, how could magic work?"

Now we come to several very interesting trends of thought. There are some people who say, "well, you see it's all psychological" and you get this. They say, "the faithful or the superstitious are in a special frame of mind where they really see these apparitions and feel these influences." All right—now nobody is much bothered about this, because psychology was the one thing they really understood. And they understood about the art of memory and about human imagination and about the power of imagination, but by 1500 they were asking themselves a very crucial question—"can this be transferred to the outside world? We are all convinced that a man can hold the audience spellbound, but will this spell work on the chairs? Will this spell work on dead nature?" And this question—"is white magic transitive—can it be transferred to dead objects?" This is what was now engaging everybody's attention. There was wonderful excitement about how different people treat this, but they all came back to the same thing. Now they took different lines.

Some of them said, "well, yes, but in very special circumstances"; or "well, yes, but it isn't exactly that you can make nature obey your will, but if you choose to do it at a moment when nature is ready, then you can just slightly distort her," and so on. Some of them, of course, began to say about this time that it's just not true. There just isn't any magic at all. For several hundred years everybody had been saying that it is well known that menstruating women must not look in mirrors because they tarnish the mirror. And then, around 1550, people said, "Have I been looking in mirrors recently? I haven't noticed any mirrors tarnishing." And of course, at that moment, all those delicious old wives' tales about the influence of man on the environment began to disappear. To summarize, I quote from Pomponazzi, who, in a book called *Of Incantations*, says quite firmly,

> It is possible to justify any experience by natural causes and natural causes only. There is no reason that could ever compel us to make any perception depend on demonic powers. There is no point in introducing supernatural agents. It is ridiculous as well as frivolous to abandon the evidence of natural reason and to search for things that are neither probable nor rational.

Well, of course, that's a very wild voice by this time. And Pomponazzi was an Aristotelian from Padua to whom these things came, as it were, from the outside; but he did mark a great turning point in this period, the time when black magic was at an end; everyone had gone through the white magic period. In black magic, the belief was that you would make nature run against her will. In white magic, you began to say, "Well you know, let's make nature work with us. There is a harmony; we could exploit it." Finally came the concept of natural law itself. And that was represented, in a most spectacular way, for the first time in the writing of Francis Bacon between 1600

and 1620. It was Francis Bacon, whom I was quoting, who
was the first person to say "knowledge is power." It was
Francis Bacon who said in the *Novum Organum* "we
cannot command nature except by obeying her." At this
point, the scientific revolution was really complete. This is
an important issue because there has been a good deal of
argument about who Francis Bacon was—whether he
wrote Shakespeare, for example. It is particularly important to determine how he fits into all this. And it's really
only since the publication of Paolo Rossi's book, *Francis
Bacon from Magic to Science* (University of Chicago
Press, 1968), that it has really become clear to us that he
went through all this; he understood all this Italian stuff.
And then at the end, he came out with this simple notion:
it wouldn't work. That's a very English thing to do. One
could have an entirely separate chapter on that Puritan
frame of mind which made him say, "all this stuff about
the face of God and the harmony of the spheres and the
number, mythology, and the love of God—how does it
really work?" At any rate, it's clear that he was outraged by
many of the fancies of the sixteenth-century writers on
memory and magic, and that he came to this crucial
conclusion, "we cannot command nature except by obeying her." There are laws of nature, and what you really do
is not to turn them back but to exploit them.

If I might give you one spectacular example, who
would have thought in 1569—when they were already
well on the way to this concept—that if you really wanted
to make the biggest bang that you ever made on earth,
you would not in any way call up the sun, call up the

volcanos, call up the mystic power; you would just take ordinary atoms of uranium and you would put the U_{238} atoms in one box and the U_{235} atoms in another box and that this simple rearrangement of nature by her own laws would blow up 120,000 people in Japan.

I've made a passing reference to a shift to England at this time and it would not be fair if I didn't draw your attention to the importance of Protestantism and Puritanism. There is a very curious history about magic I believe to be true. It has always been a puzzle why, certainly from 1640 onward and probably even before, the Protestant countries began to take the lead in science. Obviously the trial of Galileo in 1633 had a tremendous influence, but there must be something in the background of the period 1500–1600 which began to shift the center of gravity. Now I believe that this has a great deal to do with magic, for a very curious reason. By 1600 it was open to anybody to say, "you can't persuade people of one thing when another is true simply by using words." But unfortunately, Thomas Aquinas had committed himself to the statement that the words which are used in the elevation of the host have absolute power to change the bread into the body of Christ and the wine into the blood of Christ. And the statement that Thomas Aquinas made back in the 1250s was so absolute that it was really impossible to get around it. The words, "this is my body" the words "this is my blood," would, if uttered, make a difference even if they were made by a priest in bad faith, by a priest in unworthy circumstances, or by a priest who was not thinking about the subject. And if he did not utter those words, then no transubstantiation would take place. That was a very big issue throughout the sixteenth century because you could not get round the authority of Aquinas on this, and yet here was something which in some way had to be explained away, and it was very adequately explained away. One could write about Duns Scotus's view, the Thomist view, how all this was dealt with in the sixteenth century. But the fact of the matter was that it created an attitude, in my view, about the nature of science and the existence of magical powers which was different in Roman Catholic countries and in the new Protestant countries. And we know this, because the Protestant writers were busy attacking what they called the superstition of the church. And of course, in Puritan England this was especially true.

I have made this long historical excursion because I wanted to demonstrate what I think Ficino did when he suddenly opened up the world and made the rainbow full of color and said nature and man are in harmony. I said in chapter 1 that I couldn't think of any way of being a human being other than by being an intellectual. To me, being an intellectual doesn't mean knowing about intellectual issues; it means taking pleasure in them. And that to my mind is exactly what happened—exactly what transformed the attitude to science about the year 1500. The sudden sense of an opening universe—you get it in Copernicus, you get it in Galileo. If you read Galileo's *Dialogues* and all those corny jokes and all that leg pulling, here is a man who is in love with his subject and who is no longer practicing Faustian demonic magic and swearing to the devil. He is out in the open; he just thinks it's marvelous. I think, of course, that science is wonderful in that way. And I end it with Francis Bacon for that very reason—that the Elizabethan Age, to us an age of literature, was exactly that age when all this science and literature together came to fruition in England.

I quoted the *Novum Organum* of 1620. It was as late as 1620 that "knowledge is power" was written for the first time. Twenty-five years later, on Christmas day in 1645, Isaac Newton was born; in another forty years, Isaac Newton published the *Principia*, and quite suddenly the world was transformed into something which is both rational and beautiful in just the way that the neo-Platonists believed with all their Averroist and their Aristotelian tradition. Now the nature of that scientific knowledge I shall discuss in chapter 3, and its human implications in my last. But it seems right to end here by noting that what I have tried to communicate is that at one moment in history, science and the arts rose together, because of the simple sense of man's pleasure in his own gifts.

III
The Strategy
of Scientific Knowledge

I HAVE PUT forward the view that science is a world picture. It is not a technique; it is not a form of power; it is not even simply an accumulation of knowledge. But it is a highly integrated form of knowledge which makes a world view. And to this, I hold, we have been irrevocably committed roughly since the Renaissance. I hold that the scientific revolution from 1500 onward was an essential part of the Renaissance, that without it the Renaissance cannot be properly understood as a revaluation of man, as what Professor Kristeller and his colleagues call in their joint book, *The Renaissance Concept of Man.**

Since that time we have been in the unique position of trying to form a single picture of the whole of nature including man. That is a new enterprise; it differs from the preceeding enterprises in that it's not magical, by which I mean that it does not suppose the existence of two logics, a natural logic and a supernatural logic.

*New edition, New York: Harper, 1973

When the new outlook on the place of man became current around the year 1500, Aristotelian Science was, therefore, broken as a tradition, and it became necessary in some way to reestablish a relation between man and nature different from the old order of "God rules man" and "man exploits nature which has been created for him." It was impossible at that time to suppose you could do the whole thing without God, but God and man became very much equated, and the question now was how did this image, this God-man image, relate to nature? Well, naturally what began at that time was a highly personal way of looking at nature, that kind of psychological magic which I described toward the end of chapter 2, in which people no longer thought of dominating nature by a brute reversal of her laws, by black magic, but in some way by a pscyhological domination. The magus, from the time of the Hermetic books, becomes someone who persuades other human beings that they really are seeing visions, that they really are in a state of ecstasy, and who tries to think that this is a transitive gift. In their vocabulary, nature is also controlled in this way. We have the sense that nature is animated; she is that anima which Plato and Plotinus conceived, and that anima has a spirit. And man is trying to control nature as he would control an adversary—by words, by gesture, by the means that he would use if he were facing a human being.

A French scientist whom I will not name said to me once, "Pour moi, faire la conquête de la nature c'est la même chose que faire la conquête d'une femme." "For me, to dominate nature, to conquer nature, is the same kind of thing as making a conquest of a woman." Well, that's very much a 1500, white magic view; a little out of date, but that's how the French are. And perhaps in that strange phrase, which was spoken in all seriousness, you get the spirit of the relation which man was trying to establish with nature at that time. And perhaps my joke about that's how the French are is not quite out of place because this animistic view of nature really held until Francis Bacon wrote the *Novum Organum* in 1620 and put forward a view which for the first time proposed that nature was a mechanism that had to be studied and understood in quite a different way by its own laws. Of course, when I mention 1620, you must really not suppose that nobody had ever said this kind of thing before, or that on the 31st of December 1619, a new dispensation started. But it is certain that in that time a transformation took place. I referred in chapter 2 to Paolo Rossi's book, *Francis Bacon from Magic to Science*, and that the whole gamut had now been run from black magic through this white psychological magic, natural magic, to the notion "well, that's how the world is." As Francis Bacon said, "Only knowledge is power; we have to understand nature. We can no longer dominate this mistress; we can at best wheedle her by following her own idiosyncracies."

Francis Bacon has been attacked a good deal in the time that has passed since then for a number of reasons, one of which is common to most philosophers of science and in part deserved: he was never really a practicing scientist and most of the things he talked about are excessively formalized. His notion of how science should be carried out is certainly terribly naïve, as is inductionism as he proposed it. His notion that you watched nature, and you looked, and when you saw enough instances, you said "Aha! I see how it goes," and then you had a law of nature—that form of inductionism, of course, doesn't work.

For a variety of reasons we certainly are not able to match nature outside by the way we conceive of her inside our heads by thinking that we simply copy the outside world and then watch for rules which guide the instances.

Francis Bacon could be said to be formulating good rules for nature, but that's not the same thing as finding laws of nature. Let me give you a simple and as it happens a pertinent example. In my view the most brilliant piece of inductive reasoning that has ever been done was done by Dmitri Ivanovitch Mendeleev in March of 1869 when he propounded what is now called the Periodic Table— the proposal that if you wrote the different elements in the order of their atomic weight, then every eighth element was rather like the one that stood above it in the table. Now that was tremendous. First of all, it was true— provided that you didn't count exactly, you shifted the eights about a bit, you did things with the columns, and you did all those things which it always turns out you must do (science is much more informal than any of these statements about it). But a second and much more profound reason was that it turned out that this regularity meant something. It represented a real law of nature which took, of course, a long time to discover afterward— that atoms really are constructed in such a way that when finally you fill up a ring of electrons, which can consist of eight places, you start again on the next ring; and that's where the eighth-place sequence comes from. Therefore, you suddenly see a profound explanation, a law of nature which the induction in the Baconian sense had only foreshadowed. But the whole thing about the atomic table would be madly uninteresting if it hadn't led to that discovery, because who would care about having all those things written up in rows of eight and so on; there's only about a hundred of them. And if you were just making generalizations you could find fifty ingenious ways of connecting the elements so that they have some similarities. But this was a similarity, this was an induction, which really pointed to an explanation.

To my mind, all systems of science have to be thought of as systems of laws and not as systems of instances, and we must not think of a law of nature as being simply an enumeration of its instances. That's not what explanation means in science. We are not sure what it does mean; we are all busy and will be for hundreds of years still trying to refine and elucidate and illuminate this very basic human concept of what explanation means. But it certainly doesn't just mean a shorthand for a number of instances that you can step off once every eight steps. And there are very good reasons why, in fact, we could not form a systematic picture of the world which was simply based on an enumeration of nature's appearances. I can put this very bluntly by saying quite simply the fact is that the brain is too coarse to record nature in such a way that we could simply run a film of all the instances in our heads and say, "now that's how the world goes."

With all its miraculous interconnections the brain, the human brain, the most complex brain of all, is still a very coarse instrument. And it can only find regularity in nature by digging it out or putting it there (however you like to put it) but not simply by recording nature. The eye is not a mechanism which really communicates to the brain what it sees as if there were another little man in there turning a film in whose head there would, no doubt, be still another man turning another little film; nor is any of our perception of nature of that simple naïve type. Very complex processes of integration go on in the eye and in all the senses. And it is that integration which in a sense already makes laws for us, predisposes us to recognize certain kinds of law, and also makes it necessary for us to find laws if we are to find our way through nature at all.

This has been beautifully said by Alexander Pope in section 6 of *Essay on Man*. This is the section which leads up to that wonderful passage that you all know about the vast chain of being to which all the animals fit. But here he is speaking very specifically about man:

> The bliss of Man (could Pride that blessing find)
> Is not to act or think beyond mankind;
> No pow'rs of body or of soul to share,
> But what his nature and his state can bear.
> Why has not Man a microscopic eye?
> For this plain reason, Man is not a Fly.
> Say what the use, were finer optics giv'n,
> T' inspect a mite, not comprehend the heav'n?
> Or touch, if tremblingly alive all o'er,
> To smart and agonize at every pore?

> Or quick effluvia darting thro' the brain,
> Die of a rose in aromatic pain?
> If nature thunder'd in his op'ning ears,
> And stunn'd him with the music of the spheres,
> How would he wish that Heav'n had left him still
> The whisp'ring Zephyr, and the purling rill?
> Who finds not Providence all good and wise,
> Alike in what it gives, and what denies?

Well, for the philosophy of the happy mean that's superb. It says something terribly interesting. He really does say that if our senses were to give us a complete picture of the world, we would be overwhelmed in some way. At any rate, we would need what computer people call a storage capacity immensely larger than any that we possess.

Very well, science, therefore, organizes these appearances, the messages that come to us through the senses, and organizes them in some way which gives them a structure. The best analogy that can be made to that structure is to say that science is a language which has these kinds of units in it. In *The Identity of Man* I wrote:*

> Science is not so much a model of nature as a living language for describing her. It has the structure of a language, a vocabulary, a formal grammar, and a dictionary for translation. The vocabulary of science consists of its concepts all the way from universal gravity and the neutron to the neuron and the unconscious. The rules of its grammar tell us how to arrange the concepts in sensible sentences—that atoms can capture neutrons, for example, and that heavy atoms when

*New York, Doubleday/Natural History Museum, 1971, p. 42.

they split will release them. And the dictionary then translates these abstract sentences into practical observations that we can test in the everyday world: for example, in the damage that neutrons do when plutonium is split.

Now tonight I want to take this picture of the language much further. Here we have a nice simple model of a language; there is a vocabulary of the units—neutrons, neurons, complexes, forces, and the like. The way you write equations between them is essentially to write a sequence of sentences. And then what the sentences say is something that can be translated into the real world. In my view this is how human language works, and indeed originates. I am saying that science is exactly an expression of human language capacity in a special, formal way. You see, all animals have languages in some way; that is, they have means of communication. These means of communication always consist of sentences. When the dog barks at you, it is saying something fairly complex like, "Go away," or "I don't like you," or "Watch out or I'll bite." But whatever it is that he says, it's not a word; it's a sentence; it's a complete signal or message. It is what I would like to call from now on *an instruction*.

Now there are theories of human language which suppose that somehow the whole capacity of human beings to make and use language is different in kind, and that human beings really start with words—you say "Ma-Ma" and "Da-Da" to the baby, and from then on it learns to say complicated things like "eggcup."

Now in my view there are two things wrong with this. The first is that it would be very difficult to understand how the human capacity for language evolved from something which is common to all the other animals, and yet took so wholly different a shape. And secondly, we do in fact know enough now about the first things that babies say to begin to suspect that those really are sentence sounds just like animal ones, and they are meant to communicate a complete message. It is my view that we analyze the world in exactly the same way that science does, and we do that from the moment we start being the human animal. We analyze our experience in that way. At the time that we learn language, we match the analysis of experience with the analysis of what our parents say to us, and finally we analyze natural phenomena from which we control the messages of nature in that way.

How? We do something to the type of signals that animals use which takes them apart so that they can be put together again in different ways. There are many things which distinguish human language from animal language, but a crucial one is that human beings can begin with a sentence like "John loves Mary," and they can take it apart and say "Mary loves John." Then it's still a sentence, but it means something different. Now there are no animal systems that we know that do this. Zhinkin's very elaborate analysis of the language of the baboons, for instance, shows that it is precisely this which they cannot do. They have no synonyms; they have no way of rearranging things.

Is it important to take a sentence apart and rearrange it?

Yes, it's crucial. Because then you suddenly inject into the world, impose on the world, a structure which wasn't there at all. We have all lived so long with words and things and actions that we suppose that there is only one way of looking at the world. The world consists of objects, chairs, people; it consists of actions people make; and it also consists of properties and the like. But that's only our analysis of the world. A dog doesn't think of the world as consisting of trees with leaves which are green, and so on. A dog doesn't think of a man cutting down a tree; that's not the kind of action he thinks of. We don't know exactly how he analyzes the world, but we know from his visual system and from his behavior that he does not analyze it into prefabricated objects of this kind. This objectification of the world is, in my view, a highly specific human property which is, in fact, *the* linguistic property. The enormous advance that the human race has made in this terribly short evolutionary time (beginning somewhere within the last two million years) has certainly been largely based on the selective advantage of language itself and not only as a means of communication. (The baboons have a hundred signals; everybody understands the whole vocabulary; it all works very well.) The evolutionary advantage of language has not only been as a means of communication but also as a means of analyzing the environment so that you can manipulate it in your head.

Language for us is not only a way of telling somebody else something, of passing an instruction, but also of providing ourselves with cognitive sentences inside our heads. Now if this analysis of language is right, then you will see that it prefigures exactly the way in which science analyzes the real world. There is the same attempt to find constant structures in it; there is the same attempt to find action sentences which describe how something gets done, how the structure changes from this to that, the properties of things, and so on. So I am putting forward the view that the method of science, the objectification of entities, abstract concepts, or artificial concepts like atoms, is in fact a direct continuation of the human process of language, and that it is right to think of science as being simply a highly formalized language. To repeat my quotation of T. S. Eliot, from chapter 1: "Few things that can happen to a nation are more important than the invention of a new form of verse."

But what I am saying here is that few things that can happen to the world are more important than the invention of a new form of prose. Because, of course, that's exactly what happened in the seventeenth century. This is exactly where Bacon took off, and where that great Puritan tradition of science, The Oxford Group, The Invisible College, The Royal Society began. The Royal Society was preoccupied with a prose style that was plain, without embellishments. This represents a profound attempt not just to be plain and Puritan but to try to match this linguistic description of the world in precise terms. As Pope put it: "Know, there are Words, and Spells, which can control Between the fits, this Fever of the soul." And that's right; there is a great deal of human intercourse which does consist of controlling by words and spells this fever of the soul. There is a literary form which has exactly this quality, but the form that was invented in the seventeenth century in The Royal Society is an exact expression of trying to make the analysis of language itself a scientific match to how the world was seen.

I began this argument by noting that the brain is too coarse to be able to register the appearances of the world. We have to look for laws, because that's the only way the brain can work. This might be taken to be a kind of structuralist point of view. I therefore want to make it clear that that is not my view. I do think that the human brain, like every animal brain, has certain inbuilt limitations. But I do *not* think that our subsequent ways of seeing the world, and language in particular, are embedded structures. On the contrary, the philosophy that I am putting forward I would call a *constructivist* philosophy. I am saying that by this analytic point of view we make, we construct, the laws of nature as we see them. We cut up the environment into things; we make groups of things, concepts, classes, and then reconstitute them in different forms of sentences. And the whole of our imaginative process is carried on in this way; that is, our capacity to do this in the brain is built on the fact that we are able to remember in symbolic terms; we are able to attach symbols to our memory and project them forward as foresight, and in this way we are able to manipulate by imagination the environment as it might be. And language is simply a way of doing that which is communicable, but inside our heads we are using this sort of language all the time in order, to put it very briefly, to imagine. And imagination is essentially a constructivist activity; that's why I call mine a constructivist philosophy.

There is not the slightest danger of our discovering the laws of nature; there isn't much danger of our even coming very near a formulation of the laws of nature which on merely logical grounds could work in the axiomatic systems that we use. But what we do in our pedantic, pedestrian, but essentially human way is constantly to refine our inner language in a communicable form so that we are able to utter more and more sentences which make sense about nature and describe her in a lawful way. And then science tests these imaginative predictions, and so on.

There is not much danger of our discovering the laws of nature, because we mustn't suppose that we are on the brink of some kind of reductionism. We shall not find that it's all going to work out to be a lot of atoms, and that you won't have to read my words because if you simply get sent to you the total pattern of my molecular structure, you will know what I am going to say. There is no danger of that for a very simple reason—that's all metaphysics. The fact that we understand something of a machinery which matches the process of nature says nothing about whether nature is that kind of machinery. We mustn't suppose that explanation has that reductivist character. I will quote a piece of reductivism I find funny. On page 2165 of the *Oxford Dictionary*, between *theca* and *themselves*, is the word *theism*, the definition for which is: "1886. *Path*[ology]: A morbid condition characterized by headache, sleeplessness, and palpitation of the heart, caused by excessive teadrinking."

Now I have to confess that there is another definition of *theism* in the dictionary, and I don't imagine that we are anywhere near the day where the two definitions of *theism* are going to be conflated.

And by the merest chance, I have an equally nice quotation—a headline from *The New York Times*: "HAPPY CHILDHOOD IS LINKED TO ATHEISM." And that's another form of reductionism.

This constructivist view of how science works is, of course, essentially a very modest view, and I have stressed that it is an extension, a formalization of a very straightforward kind of human activity. I have said that we must not regard it as a metaphysical statement about how the world is going to turn out to be in the last analysis, because it's quite certain that no axiomatic system can be the last analysis. Then how are we to regard it? Is any system as good as any other? No. We must regard this as a strategy for exploring nature, and that's why I call this chapter "The Strategy of Scientific Knowledge." I think it's a grave mistake to think of the progress of science as other than a workmanlike strategy for dealing with the phenomena, understanding them, explaining them, getting an enormous pleasure out of this intellectual exercise, and at the same time, as Bacon predicted, generating power from it because we unleash more of the hidden potential in nature as we deepen our understanding.

William Kingdom Clifford in the last century wrote a book called *The Common Sense of the Exact Sciences* which I admired so much that I called my first book about science by almost the same title.* And this quotation seems to me as true today as when it was written by that extraordinary, brilliant, erratic, and neglected man,

> Remember, then, that scientific thought is the guide of action; that the truth at which it arrives is not that which we can ideally contemplate without error, but that which we may act upon without fear; and you cannot fail to see that scientific thought is not an accompaniment or condition of human progress, but human progress itself.

I shall come back to the last phrase, but let me concentrate first on the central content. "The truth at which scientific thought arrives is not that which we can ideally contemplate without error, but that which we may act upon without fear." And that's crucial. Science has, of course, prospered only in those nations in which action has been regarded as a higher good than contemplation. And the sense that contemplation has to arrive at a truth for ever and ever is wholly alien to the view of science that I am putting forward, which is an activity to which we are guided by our knowledge, and which will turn out to be fallible, and then we'll devise new laws, and so on.

As for Clifford's happy claim that science is not merely an accompaniment or condition of human progress but human progress itself—well, he was trying to say much the same thing about the progressive nature of scientific

*Common Sense of Science (New York: Random House; paperback ed.: Vintage, 1959).

knowledge; he wasn't getting into any arguments about whether there is real progress and whether we are any happier. He was making the point that science is a progressive activity; it is the codification of that essential human characteristic, the making of plans. And it is so always on a highly experimental, highly tentative basis.

I shall discuss plans and ethics in detail in the last chapter; here I want to draw your attention only to one central thing—in the view that I am putting forward, science is *not* a problem-solving activity; it is *not* an improvised, moment-to-moment activity which faces us with a problem that we solve, and then go on to the next one. And though I have a great admiration for my friend and colleague Karl Popper, in his recent work he has begun to stress the notion that there is a great problem-solving element in making laws of science. I think he suffers, as so many of his colleagues do, from the fact that he really isn't used to how a laboratory carries on. There aren't any clear-cut problems; there certainly aren't any decisions in which you set up an experiment and you say, "Here's a law, here's a hypothesis, I challenge it, I'm going to negate it." Instead, it all works by a highly tentative and experimental process.

Professor Chargaff has made enormously important findings on which, in one sense, the whole of modern molecular biology is based; namely, that there are equal amounts of one base and another base, and then equal amounts of a third base and a fourth. All that was discovered not by setting up a master hypothesis and not by setting up a great single experiment, but in a series of papers in one of which he says, "Well, it is interesting to observe," and then in the next he says, "It may strike one as an odd coincidence" and so it goes on you see until finally it comes out and it is so, and the lock is put on it. At this point, a hypothesis is formed about the structure of DNA which suddenly illuminates, so that the theory and the experiments each support one another because they make a kind of jigsaw puzzle of sense.

Now it is in this sense, this highly experimental sense, that I have been describing science, exactly the sense of learning a language, of enlarging a language, just as we are doing today. If we were to go through the 5,000-odd words that I have written in this chapter, we would certainly find at least a hundred locutions or words which would not have been in common use fifty years ago. And this is as much an enlargement of our culture, as much an enlargement of understanding, as any scientific discovery.

Popper and I share the view that the process of science does consist of forming a hypothesis, of drawing some logical deduction from it which can be subjected to test, and of subjecting that deduction to test. But that is altogether too formal a view as are, indeed, almost all accounts of scientific method that have been written by anyone—philosophers and nonphilosophers. Because just as it has turned out to be enormously difficult to write about the development of language, so it turns out to be very difficult to describe this extraordinary jigsaw puzzle of activity. We don't really have any kind of language which doesn't have that nice linear form by which you go from A to B to C and so on. And that, of course, is not how intellectual discoveries are made. Above all, it is not how those great imaginative leaps are made by which new hypotheses are formed.

How do the scientific laws come into anybody's head anyway? Now we are right back to magic and the sixteenth century. In the middle of all this controversy about spells and words and so on, Erastus, the great Protestant critic, attacking everything that had been written about natural magic by others, attacked, in particular, the conception that one really could have imaginative domination even of human beings. He got somewhat carried away by this argument: "Certainly no one in their right mind will think that an image fashioned in the spirit of my fantasy can go out of my brain and leap into the head of another man." Well, it's nice to think that Protestant theologians thought that you couldn't communicate imaginative ideas. But, of course, the fact is that the whole of human language would fail if you couldn't do that, and certainly the whole of science would be hopeless if you couldn't do that.

The exciting thing about the human situation is exactly that for us there is a shifting boundary between the inner self and the outside world, in which we learn enough about the outside world to make suppositions about it which suddenly have a tremendous imaginative force for us—so much an imaginative force that when you really see a scientist who has a good idea, you understand for the first time that for him it's exactly like a poetic image. I have actually seen a man; I remember to this day where it happened, where he stood, at a time when I was evolving an elaborate theory about how poetic imagery worked. I saw him look at something and hear what somebody said;

then his face lit up and he said, "But if that is how the imaginal disc goes, then it should do this, that, and the other as the insect grows up." And you could see the whole of this unfolding. The particular fact which someone proposed to him suddenly contained the general.

Now we don't communicate science in that way; we are very careful to discuss generalizations with generalizations. We still carry with us all these nice animistic words about nature—force, gravity, stimulus, and all those—which are supposed to be neutral enough now for generalizations to be carried from person to person. But to the man who makes the discovery, the particular fact has exactly the force of the poetic image—exactly as at the moment when you hear William Blake say,

> A dog starv'd at his Master's Gate
> Predicts the ruin of the State.

And you say, "My, God, the dog starved!" Suddenly, this tells you all about a civilization which is going downhill. And just as that particular carries for all of us this great general conception, so for the scientist making the discovery, the little particular observation about how the weight, the masses, the bases, and so on carries that force. And it is in such moments that the human imagination works. How? Because we are able to manipulate in language or other symbolism, inside our heads, unrealized situations. Imagination is the manipulation in our heads of images into futures which have not happened (I am using the word *image* to mean any kind of symbolic structure like words, for example).

Now that's how hypotheses are made, and they are then tested in other particular cases. But it's hopeless to think of this as an explicable process or as a formalizable one. As a matter of fact, we don't even know how to go from one hypothesis to another.

It's fairly easy to say all this today, because we've lived for fifty years now with a wealth of change in the world of science which would have been inconceivable a hundred and fifty years ago. I once mentioned this to one of my philosophic colleagues at Harvard, and he said, "Dear man, well perhaps it will just last a little while; perhaps science will sort of settle down soon." Well, I don't think it will.

Why have we suddenly become aware that no explanations are ultimate explanations? For very simple reasons. In 1900 everybody knew that Newtonian physics didn't work in certain instances, but nobody dreamt that you would have to conceive things so differently that you would not talk about force anymore, that you would have to make a new concept of mass, or that the law of inverse squares would not hold. Philosophically, a wholly different picture now exists. When Newton says in the third book of the *Principia* (second edition) roughly speaking "I am only going to make one central hypothesis and that's that the center of the world is fixed," that's exactly the hypothesis which came to an end in 1905 when Einstein invented relativity. A wholly new conception came into being. And then we all clutched our heads and said, "That's unbelievable, you know. Here we've been going since 1687, and we have been sure that Newtonian physics worked so well it couldn't be wrong; it *couldn't* be wrong." Just like that day in 1666 when Newton sat in his mother's garden and he calculated the orbit of the moon under the supposition of the law of inverse squares, and the orbit came out at twenty-eight days. He didn't muck about anymore, and go around saying, "I wonder whether it's true," he just said, "twenty-eight days can't be an accident; I must have the law right." And so we felt in all that time since then that Newtonian laws must be right because it couldn't be an accident that you could write a nautical almanac and foretell when there would be a tide in Tierra del Fuego, and then be told, in 1905, that the whole thing was conceptually utterly false. Now is conceptually utterly false so important? Yes, because that's really the only interesting thing about it. I am all in favor of calculating the tides in Tierra del Fuego, but I do not regard it as a major intellectual activity; whereas, I do think that understanding the world—having a picture of why the figures come out in the world—is a major intellectual activity. I am right back there with Giorgio and the figures being the face of God.

In the end the calculations owe their power to exactly this intellectual beauty. But what makes them beautiful? Their unity, the fact that we have a picture of the world. And when a picture of the world, when a set of axioms, can change as rapidly as it did in 1905, we all suddenly become extremely humble about the ultimacy of any scientific truth. We are conscious that we are engaged in a strategy; that it is an unbounded, progressive strategy; that we are not, as it were, foreseeing some future problem and solving it. We have formed a policy of strategy, of doing research in a certain way, of believing how the laws go until we prove them otherwise. And it is that sense of the strategy of scientific knowledge which has been responsible for the triumphs of the last three hundred years, and for the intellectual pleasure which they give me, and I hope you.

In the next chapter I shall note that if scientists have a modest view in deference to the basic Platonic humanism from which they started in 1500, then it behooves humanists also to show a certain modesty.

IV
Human Plans
and Civilized Values

IF SCIENCE IS indeed an irreversible step in cultural history, if a change in kind has taken place somewhere about 400 years ago, does any place remain for ethics and values? That question could be asked on two different grounds. First, one could say, "it seems that all our actions will ultimately be explained in the mechanism of physical structures. And if that is so, then there is no freedom of action and it's pointless to ask should I do such, should I behave well, should I go to college, should I learn anything? In a sort of Calvinistic way it's all preordained, so why should I bother?"

In chapter 3 I noted that this is a metaphysical speculation, and that like all metaphysical speculations it really has no influence on action at all. It's quite a pointless analysis. It may well be true, as Laplace thought, that the Universe unrolls itself in some imperturbable way, with perhaps a few quantum oscillations; but by and large nothing outside that physical sequence matters. And in particular it may well be that we are under an illusion when we think that learning to read is anything but a mechanical response to the way we were made at birth and by our subsequent growing up. It may be that all that talk about nature and nurture, about instinct and plan, is beside the point; perhaps they're really all the same thing, and all plans, all human thoughts, are only *Schein* or epiphenomena—only appearances.

I call this a rather pointless piece of metaphysics because it doesn't tell you anything at all. When you've said all that, it has absolutely no influence on your conduct. You labor under this profound belief that the reason why you want to be a well educated man and get A's in all your subjects has to do with the way you were made. But it doesn't make any difference—you behave in exactly the same way, you go to the same lectures, you take the same exams. In other words, the notion that there is an analysis into the pure mechanism of our lives is a piece of unrealistic speculation exactly like solipsism. Now everybody knows that solipsism can't be disproved; you cannot prove to anybody that other people are real.

In a classical joke which Bertrand Russell made to me, and I'm sure to many other people, he said that he loved the lady who wrote to him saying that she was indignant that since solipsism was so obviously not capable of disproof, why did more people not believe in it? I suppose that one should then say to her, "Why can't you imagine more people believing in it?" But we are really in the same boat about the metaphysical notion that all life might be reduced to some kind of predictable mechanism, because it's really neither here nor there; no one would behave in the least differently. If you come to a lecture, it is because you think you want to come, and whether your thoughts are a fantasy or reality, you come just the same. Now I think this is a very powerful point which has largely been missed in all the excitement about predestination and free will and so on, but I won't labor it.

However, there could be a second reason for saying, "If science has now become *the* world view forever and forever (that is, it can only be changed in a forward direction; we can't go back) perhaps there is another reason for not wanting ethics and values." And it's this: if the world has that kind of machinery, then it doesn't have any meaning, life doesn't have a plan, there isn't anything that man was created for; and in those circumstances you might say quite seriously, "Ah, well, now I really am happy to behave badly, now you can do what you like with your lecture, I'm not interested, I was only interested when I thought that being a human being was some fulfillment of some greater purpose. But if there is no plan, if there is no purpose, if the whole universe can be reduced to a piece of machinery, well then you really can't persuade me to come. I shall go and booze it up outside."

You will notice that I have used the same word *plan* that I used in the first example, where I asked, "Can we plan our conduct?" In the second I asked, "Is there a plan to the universe?" In my view, most discussions about the second question fail by not understanding the nature of human plans.

It is sometimes thought both by poets and technologists that unless one makes a distinction between the mechanism of life and deeper truths, no system of ethics can be founded. Hume said that no amount of science will derive values from facts; it was restated as the naturalist fallacy by G. E. Moore. Max Weber gave a great deal of currency to this discussion when he said, "you can't get values out of science, but you must practice it with passion and that makes it valuable." But all these ideas contain the supposition that values must have a different kind of character from the ordinary activity of life. Now this sort of dualism is what I call *magic*; I inveighed against it in the first chapter; I've been carrying on the crusade in the subsequent two and I want to inveigh against it again here. It is still present with us all the time in that Cartesian dualism which tries to distinguish between the soul and the body as if they were really separable entities. The point of view that I have put forward as being one of the five central points about science (two of the five) is that only a unitary view will do, that any attempt to bifurcate the activities of life and to set up a dualism becomes automatically a kind of magic.

In chapter 1 I quoted a poem by e. e. cummings. Let me now quote one by W. B. Yeats. It comes out of *The Tower* and consists of two fragments. The first fragment says,

> Locke sank into a swoon;
> The Garden died;
> God took the spinning-jenny
> Out of his side.

And the second fragment asks,

> Where got I that truth?
> Out of a medium's mouth,
> Out of nothing it came,
> Out of the forest loam,
> Out of dark night where lay
> The crowns of Nineveh.

Well there you have the spectacular distinction. The first fragment says Locke is representing philosophy, science, everything that Blake called "Locke, Newton, and unbelief."

> Locke sank into a swoon;
> The Garden died;

Garden with a capital G, Garden of Eden.

> God took the spinning-jenny
> Out of his side.

It's extremely heavily weighted in favor of saying that the Garden of Eden dies and up comes technology. Now I have trouble enough with technologists who think that they are magicians, but now you see the second verse is much worse because who is the magician? The poet.

> Where got I that truth?
> Out of a medium's mouth,
> Out of nothing it came,
> Out of the forest loam,
> Out of dark night where lay
> The crowns of Nineveh.

Now I think that's a terribly destructive point of view. It's perfectly true that Yeat's wife was a medium, and that she did produce this kind of message; she certainly couldn't produce a spinning-jenny. But the notion that this great truth about how the garden dies comes to you by some kind of prophetic inspiration, that that's the only way that you get at what ethics and values are about, I think to be absolutely fallacious. And equally fallacious is the notion that the technologist is another kind of initiate. What's wrong with all these magical views is that they somehow set up a world in which there are people who know and the rest of you chaps down there—you listen, you go away, you say, "Marvelous man, how does he do it?" But you don't do anything about it, you leave it all to me. Well, you see, I do not think that the world is made of human beings who are any different from me. Now Yeats, I think, did think that the world was made of different human beings, but since he's dead we ought not really to argue about that.

If there's any reason why the magical point of view must be broken down, it is because the essence of the dualism is that there are magicians; there is a Magus, there is a comforter of the spirit or the soul, in the Cartesian terms, who is different from an ordinary human being dealing with daily contact. There is an iniate or what Calvin used to say "an elect," which can really do anything, because the elect are like those tall handsome engineers in white togas who used to walk around in the books of H. G. Wells. The engineers go around and, blessing with a papal gesture, say, "My good people be happy, we will look after your comfort, we will see that the space ships run on time—or that the trains run on time." After all, what else did the whole of Mussolini's empire consist of? Now I think that neither the space ships nor the trains are what life is about, and I think that if you adhere to this view you put yourself in the position of the kind of slave society that believes that there are elect, that there are initiates, that they know. The doctrine that technology dissociates itself from ordinary human issues, sets the elect aside, and says, "We will practice a mighty paternalism" is basically a false one, for reasons which I shall develop.

First, however, permit me to digress for a moment. There is one thing you do have to forgive technologists— they love what they are doing. Max Weber was absolutely right when he said that in the end every human being has a passion about what he does. And very often the reason why they feel that other people are inferior is they know that other people work in banks, or wash dishes, or remain at home as the baby-sitter—things which they don't love doing and don't have any passion about. I think we are very fortunate to be the elect in doing things we really like. I once said in an interview on the BBC that I have had a marvelous life because I've always been paid to do what I liked—just like a prostitute. The BBC left out the last phrase. But, of course, technicians love what they are doing, and therefore, for instance, it is quite certain that all those people who worked in Los Alamos were going to blow that bomb; you couldn't stop them. But to say this is not at all to say that they were going to blow it and kill 120,000 people. That's not what the Wigner-Szilard memorandum proposed. They were, of course, madly anxious to demonstrate the bomb and, of course, they were full of excuses about how it would be good for people to see it, when really they were just dying to see it go off. But always remember that dying to see it go off carries all the excitement of flying a rocket to the moon, and has nothing to do with the excitement of killing a great many people.

So you must allow in all activities—Yeats's as well as the technician's and the scientist's—the sense that skill does give you a tremendous urge to see your work made to work. As a matter of fact, it's very important, because one of the things about the idea of the elect is that they love to work in secret.

A French scientist worked for me during the war, and he would always say when he saw a paper which was marked "Secret," "Je pense qu'il existe une petite erreur" (I think there is a small error). But when he saw the new classification which the Americans brought over which was "Top Secret" then he said, "Je sais qu'il existe une grande erreur" (I know there is a large error). Now he might not have meant an error in the sense of just a mathematical error, but you see basically this is terribly true, and the whole point about secrecy and science, secret research of every kind, is connected with this. Why do scientists keep on saying, "We don't want a moratorium, we want all kinds of things, but you really must make discoveries"? It is simply because with the coming of secrecy the whole scientific tradition became distorted.

69   Human Plans and Civilized Values

The ability to make progress in a natural direction rather than to solve problems in a predestined direction would go. This is a large issue, and I cite a little poem by Robert Burns which is really about secrecy in love affairs, but it's just as true about secrecy in science:

> I waive the quantum of the sin,
> The hazard of concealing;
> But och! it hardens all within,
> And petrifies the feeling!

Now everybody knows that that's true about an illicit love affair, which is a very fortunate state of affairs. But it is also essentially true about the practice of science and why secrecy in matters of research is such a fundamental evil, and why therefore the pleasure in wanting the research to come out, the wish to be published, the anxiety not to work in establishments bound by patents, rules, and laws is such a very healthy part of scientific progress.

Since I have said so much about the concept of the elect, and since I made some remarks about Roman Catholicism and magic which might make one think that I'm prejudiced in religious matters, let me just underline the Calvinistic message by reading from Erastus, a great attacker of church and magic, who believed that Protestantism would do away with it all. In chapter 3, I cited a passage in which he said the imagination couldn't exist— that it couldn't leap from one man's brain into another. That was the kind of box that he had got himself in. Here he is saying something about magic and particularly about formalities and ceremonials:

> There is no power in ceremonies but that of representing. For ceremonies have been instituted for the sake of representation, or indeed of order and splendour, so that, striking the eyes of the less educated, they might help both the understanding and the memory.

D. P. Walker, from whose book on magic I am quoting,* has a dry footnote about the passage, saying that it seems to imply "that intelligent Christians with good memories do not need any ceremonies." But I'm not making a joke about it, I am simply showing how easy it is to get into this notion of the elect, that you are in a special category, that your kind of knowledge is in a special category. And I think this to be fatal whether it is said by poets like Yeats or by technologists who think that you can solve the problems of the world simply by their special kind of technique. Very well. If we then are to escape this, on what can we found the idea that a developing ethic is indeed a proper human expression? Obviously we must found it in the specificity of the human species. We must say that planning is an important part of the way that the human mind works. This is an important subject— whether or not you believe that it's all done by molecular biology. Those are the facts; this is the kind of plan—and it is different from the behavior of other animals. Human beings have these special characteristics of memory, imagination, symbolic representation, and language which make it possible for them to project themselves into futures which have not yet happened (and, indeed, some

*Spiritual and Demonic Magic from Ficino to Campanella (London: Warburg Institute, 1958).

of which will not happen), to set up artificial futures and make plans toward one rather than another. This is the central act of choice, and it's really quite irrelevant whether there is or is not free will, because that's how people behave. And just as the behavior of other animals is studied in terms of how they behave, so the behavior of human beings has to be recognized as what it is. What is very special about human beings is that in making their plans they look into not only the outside world but also into themselves. The paradoxes of science and the paradoxes of literature both derive from this ability of the human mind and of all human language to frame self-referential questions. Now this subject about questioning oneself, looking inside oneself, is one which has fascinated both scientists and poets. One poem which I think is a superb discussion of this problem is the eighth of Rilke's *Duino Elegies*.

Rilke was very much interested in this question about the inward-looking man and the outer-looking animal, and his friend Rudolph Kassner, who held a quite different opinion, was also interested in this question—the openness of the world to the straightforward animal vision, and the tendency of human vision to return on itself. I quote a few lines from the translation by J. B. Leishman and Stephen Spender of the Eighth Elegy which is wholly concerned with this theme.

> With all its eyes the creature-world beholds
> the open. But our eyes, as though reversed,
> encircle it on every side, like traps
> set round its unobstructed path to freedom.

> What *is* outside, we know from the brute's face
> alone; for while a child's quite small we take it
> and turn it round and force it to look backwards
> at conformation, not that openness
> so deep within the brute's face.

It took two poets of some distinction to write that translation. Unfortunately Rilke said nothing about "the creature-world beholds." He did say:

> Mit allen Augen sieht die Kreatur
> das Offene. Nur unsre Augen sind
> wie umgekehrt [as if inverted] und ganz um sie gestellt
> als Fallen, rings um ihren freien Ausgang.
> Was draussen ist, wir wissens aus des Tiers
> Antlitz allein; denn schon das frühe Kind
> wenden wir um und zwingens, dass es rückwärts
> Gestaltung sehe, nicht das Offne, das
> im Tiergesicht so tief ist.

Now, of course, Rilke didn't really mean that what is outside we only know from the face of the animal, because we know it from ourselves also. But what he meant was that in a way the animal's face reveals what's outside much better than our own, because the animal does not have these self-reflective powers. We know from its shortcomings of memory and the like that it does not have the symbolism to turn these questions on itself. Now these are very important issues, to which I shall return. Here, I will say only that they show us that knowing about the physical world is not simply a process of experimenting as if the world existed outside ourselves. There is no language which describes the outside world like that, although science, of course, tries to do it all the time. And yet it tries to do it in terms in which other people are also accessible to our kind of insight. We have the extraordinary and unique gift that we are able to see ourselves and to recognize ourselves in others and then to recognize the human condition in ourselves. Now this is an immense marvel. In some way you, reading my words, are not simply reading a stream of messages; you're reading the words of someone who is like you. You recognize in me foibles of your own—little things which you hide from yourself, little vanities which you don't reveal to anybody else but you are amused to see that I also display; and suddenly you say to yourself, "That's how human beings are." And at the same time you recognize not only that everybody is to that extent like you but also that humanity has these multiple faces of which your particular arrangement of facets is unique—just as your body chemistry is unique, although it's simply made up of a different arrangement of animal elements that we all share. So that when we go to lynch a murderer, it's not because we hate murder, but because we're so afraid of the murder in ourselves that in the end we commit it.

And it's this sense that the human being is all of us and yet recognizable in ourselves which gives that part of our analysis such a different character from a mere study of the world outside ourselves. We are part of the world in the sense that we have to analyze the experience, and that's just our way of taking it apart and putting it together again.

One of the things that we understand, therefore, is the way in which human beings plan. And particularly we understand that there is a conflict of values in every plan that we make. The evolution of the planning activity is one of the great distinguishing things in human beings in the same sense in which I talked of language (chapter 3) as being highly species-specific. If we are to find a morality and ethic and understanding of the world, a sense that the world makes sense, we are bound to find it in the fact that we as human beings interest ourselves. We find that exactly by the kinship with other people and back into ourselves. How did this come about? In some way over a period of around two million years we have learned to walk upright, thereby to liberate the hands and, particularly when we came down from the trees, to liberate the hands even from the function of tree climbing. The liberation of the hands leads to the liberation of the mouth, because once you can begin to manipulate food with your hands, the mouth no longer is preoccupied with grabbing and biting and masticating food. Now the mouth becomes free for other things, of which speech, of course, is easily the most important, but which also sets up in us the face-to-face contact on which human life is based.

Long before this, the eyes had already rotated forward into the head, as they have in all the primates, so that we have binocular vision. But now a great increase in musculation took place, around the mouth and around the hands. An enormous amount of the nervous system in human beings is concerned with the manipulation of the mouth, and this brings with it of course not only the manipulation of things like speech but also all face-to-face contact; human beings carry out all their activities in face-to-face contact. For instance, we are the only animals that practice sexual intercourse face-to-face, and do so in all cultures; this is the basic human mode, just as this is the wholly abnormal mode in those few animals like the higher primates who do experiment with sexual posture. The mouth becomes an erogenous zone, and with the development of speech and of the hands there grew up, probably in the last half million years, a great efflorescence of the brain.

So far as we know, about a half a million years ago the brain weighed about a pound (as it now does in a chimpanzee), perhaps a pound and a half; now the average human brain weighs about three pounds, so it has at least doubled in size. And this is not just a question of doubling in size, but in very specific points. There has been a great increase in the density of cells in the visual system, so that we see much better. And that, of course, is one reason why we are able to analyze the world into objects in a way in which a dog never could. Part of the brain became devoted to speech—a part closely associated with the visual centers—and a great part of the brain became devoted to the manipulation of the hands. The bifurcation of the brain, which is characteristically human, occurred, so that we really for most purposes use half the brain. Handedness, which is absent in other animals, developed, and the great frontal lobes evolved—those which make us high-browed and those which make us eggheads in some way they coordinate memory and imagination. We really don't quite know what the frontal lobes do, except that damage to them is very destructive of the whole integration of the personality even in little monkeys. Little monkeys who have damage to their frontal lobes can do almost any kind of mechanical task just as well as before and can even learn it, but as soon as, for instance, you show them an object and hide it they're back to those days of the first months of a child's life where out of sight is out of mind.

Now the apparatus as I have described it is much concerned, of course, with manipulation and tools, but it is equally concerned with the ability to take the past into the future—to use hindsight (a great and central human gift) and project it into foresight. And that's why I say that the activity of planning is certainly basic to man and probably the great cultural pressure in human evolution has been the choice of mates, with good manual dexterity and good speech, good ability to make plans in quite primitive communities. And I am really now thinking of *Australopithecus*, an extremely primitive little order of primate. Now I have used the word plan throughout this discussion to indicate foresight and hindsight, and before I go on to a deeper discussion of plan, I should make one more remark about hindsight. If you are to ask of the plans that you make whether they succeeded or not, then you must have some criterion which is equivalent to saying, "Did it happen like this, or did it not?" Now this criterion of objective truth, of making cognitive statements about which you could say they are true or they are false in a purely objective way, is a quite characteristically human thing. I cannot think how you could devise an animal experiment in which you could ask of the animal "is this true or is this false?" And of course it's very characteristic that all the paradoxes of human logic occur only when you begin to introduce the words true and false. I think that true and false are very primitive notions to do with the match of prediction to performance, with the ultimate test of hindsight, and therefore have a great selective advantage in human evolution.

Some of the questions to ask about this sort of description of evolution are "Where is it going?" "Why did it happen?" "What was the plan?" And the terms in which I have described it to you show that that's not a sensible question. Nothing about what I have said supposes that there was a planner who made that plan. Nothing that I have said supposes that in our evolution the human race is solving some cosmic problem. You see, science has tended to get connected very much in people's minds with a sort of favorite engineering phrase, "the problem-solving capacity," or "the problem-solving activity." But problem solving is an extremely lowly form of imaginative activity. Problem finding (finding real problems) takes some doing, but problem solving is not what planning is really about. A rather attractive example of problem finding is Pascal's wager on the question of how one should behave about the hereafter. Pascal said it's really a very simple problem: If God exists, then if we don't behave properly we shall be damned; if he doesn't exist, it doesn't matter anyway. So we might just as well accept this wager and behave as if God existed. It's a sort of curious phenomenon that God is somehow not quite as nice as the devil; the devil doesn't punish you for behaving well, but God punishes you for behaving badly. Pascal's wager is a one-sided wager because it's built on the notion that the devil doesn't ask for cash. But let us accept Pascal's wager. What strikes one about it is that if it is really an ethical plan, it is unutterably contemptible. Any deeply religious person would have said to Pascal, "I would rather be damned than believe on those grounds," and it's no wonder that Pascal was so deeply divided and unhappy a man if that is how he really felt. And why? Because he thought of ethics as if it were a question of problem solving.

Now let me take another example, that of Jim Watson's instructive and delightful book, *The Double Helix*. It is a book about how he and Francis Crick made the discovery of how DNA is constructed. Now people say all kinds of nasty things about Jim Watson and *The Double Helix*; they say how could a man go around thinking about the Nobel Prize when he ought to have been thinking about the higher ends of science? How can a man go into research in this highly competitive spirit? And by and large they persuade themselves that they are nicer than Jim Watson, which is what they must believe since he won the Nobel Prize. But the crux of the matter is that the book is so exciting because it reveals, in a very plain way, what the nature of that kind of plan really is. Here's a man who plans to do something great in the world. He doesn't go out like the chaps in Truman Capote's *In Cold Blood* and kill people. That's a very simple solution to the problem. You want to make the headlines, take a gun, strangle eighteen nurses, and so on.

But Watson really would like to win the Nobel Prize. He doesn't start one of those B movies, you know, with the great plot about how by hanging upside down in Copenhagen you can steal the Nobel Prize. What he conceives is a plan of life leading toward this great achievement, and the first thing he does is to find a worthwhile problem. And when he has the problem, then (you can see his nostrils dilating) he says, "You know, I am really on to something; by Jove, I really might win the Prize." And you can't blame him. But then all that sort of sense of pursuit in the book, of going for an objective, is because there is this deeply underlying plan that "I have a grand problem, this is something very well worth doing, and I will organize my life to achieve it."

Well now, as a scientist, how do you organize your life to achieve it? One of the things you don't do is to steal a lot of other people's work, because the plan of science is that you don't win anything with stolen work. And if you look into the whole of the way, a very simple, very delightful, terribly naive person emerges (the wonderful quality of the book is that it's written by somebody who does not wear his heart on his sleeve, but actually walks about in sleeveless shirts), with a transparent sense of how you get at this. He tells you that you plan an activity; you go off and you read all kinds of books containing anything that seems relevant; you ask all kinds of people. Of course, you're constantly feeling that you're not very attractive and that you'd really like to go to the cinema; but all the time there's this consolation, "All right, so I don't get any dates, but I am going to solve this problem. And I've got a good problem, a crucial act of judgment."

Now I call plans of this kind unbounded plans, by which I mean that they are not plans which are directed toward a specific objective whose solution is obtainable by a known number of steps. They are unbounded plans, because you can devise a strategy for getting at them but you cannot devise a series of tactics; and in my view, the great feature of human behavior is that we have strategies by which we direct our lives. And those strategies are intended to achieve something; we couldn't easily put it into words. We couldn't say very easily, "Well, you know, I'd like to be happy and quiet, and I'd like the children to be better behaved and I really would like not to have to be divorced yet again." All these are things on which we center our lives, but all of them amount to saying that there are certain general ends which we are trying to achieve, like solving the DNA structure. And we do not know by a tactical set of steps how to solve it as if it were a problem. All we know is that we believe we can form general rules of our own conduct which will make this kind of unbounded plan. And the name of such an unbounded plan is *values*. We simply say, do you want to lead a happy life? Well, we should set a value on friendship; we should set a value on a certain kind of abstinence; we should set a particular value on treating other people as we would like to be treated, and so on. It's pointless my going on, because it's all said much better in the Sermon on the Mount with which I am not competing. But it has this central character that an open plan, an unbounded plan is directed toward achieving general goals in life, in science, and it is not a series of solutions of problems. It does not take Pascal's way out by saying, "I know a trick for solving that problem."

These open plans, unbounded plans, depend on our recognizing that other people and we are one—that we understand them through ourselves and understand ourselves the better through them. And for that reason it is basic to any ethical strategy of life—to a scientific strategy such as I described in chapter 3 or to an ethical strategy for carrying on one's life. And I am not using the word strategy in the pejorative sense; I'm just using it for a long-term view of how one wants to come out of it. It's central to those strategies that we should recognize that all human beings are human in just the sense that we are. And that's why I am so much against the view of the Magus, the initiate, the elect, because there aren't any such people. I don't know any such people, the most that I can say about other people is that I recognize myself in them and them in me.

Now this deep sense of human kinship is I believe basic to all human values; it doesn't matter whether you are involved in science or anything else. You can't in fact carry on a scientific activity without just that sense of truth and dignity, that sense that this human relation is a means, and there can be no higher ends. Now if you don't believe that, then, of course, you fall into the kind of trap into which Nazi Germany fell. The Nazis were in a very curious paradoxical position, because on the one hand on general cultural grounds they believed that power must come from knowledge. This is a notion which has been ingrained for the last 400 years. But on the other hand they didn't really believe in this idea because of the particular aspect of power and Arianism and the elect that they had. So they were in this strange bind of wanting to have research done and wanting knowledge, but feeling that they had some right to power other than knowledge.

Much of the Nazis' viciousness toward prisoners, Jews, and Communists is a kind of hidden anger about the notion that such non-Nazis could also be human beings and therefore be entitled to humane treatment. Remember, one of the most fundamental things about the concentration camps is not that people were beaten and tortured, burned and killed, but that there was a deliberate attempt to degrade and dehumanize people from the outset. For instance, it was a simple piece of policy in a concentration camp that the women's lavatories and the men's latrines faced each other across a wire in the open and that you were not allowed to perform your physical functions (which often, in the conditions of imprisonment, with that kind of food, were extremely distressing) except in full sight of people of the opposite sex. Now you might think that this is a kind of refinement of inhumanity, but it's really simply the basic inhumanity: the same sort that we resorted to when we went to war with the Japanese, and at once painted them so that they looked like rats. We could thus get rid of the notion that they are also human. Now the sense in which that's most displayed is in human experiments, and it's about those that I want to talk finally in order to talk about the ethics of science.

People sometimes write as if a lot of scientists worked in concentration camps, just as they often write about the

German love of culture and music. The scientists listened to a symphony and then went out and tortured people. I can assure you that the concentration camps were not staffed either by musicians or by scientists. They were staffed by thugs, some of whom, of course, were fond of music and some of whom even had some ability in science. But let's concentrate absolutely on the medical part of it. I'll read you one verse from Bertolt Brecht's "Furcht & Elend des III Reiches." *Fear and Misery in the Third Reich:*

> Es kommen die Herren Gelehrten
> Mit falschen Teutonenbärten
> und furchterfülltem Blick
>
> [The scholars come with false Teutonic beards
> and eyes filled with fear.]
>
> Sie wollen nicht eine richtige
> Sondern eine arisch gesichtige
> Genehmigte deutsche Physik.
>
> [They are not interested in genuine science so
> long as the science is Arian and bends itself to
> that kind of ideology it will have to to.]

Well it's a perfect description of the kind of experimentation which in fact took place under the Nazi regime.

In 1946 there were held the great trials of 23 German doctors, of whom seven were hanged, and some of them were men of some distinction in science. For instance, Karl Gebhardt was the president of the German Red Cross. He was also personal physician to Himmler, and it happens that some of the experiments he supervised had a very special character, because when Heidrich was shot down in Czechoslovakia, Gebhardt was called in and tried to cure the gunshot wound without using sulfa drugs; Heidrich developed gangrene—gas gangrene—and died. And Himmler said to Gebhardt, "You'll never persuade me that you shouldn't have used sulfa drugs until I've actually seen experiments carried out on other people." And so he went to the concentration camps and carried out the experiments, and since the misfortune was that the sulfa drugs worked quite well, it was necessary for prisoners to die in order that Gebhardt should save his reputation and his life. Now we must treat that kind of case as in a category of its own. But if you look down the whole of that trial of these 23 doctors, you will find that all their research is directed to technological subjects like this business about sulfa drugs and gas gangrene. The account of the trial is published in English, as always under a ghastly title, *Doctors of Infamy*,* but it was written by two Germans who went to the trial, Alexander Mitscherlich and Fred Mielke, under what I think is a very splendid German title, *Das Diktat der Menschenverachtung, The Dictate of the Contempt for Man*. That's what it's about, and that's how Germans should regard it. Now if you look in the index of the book, you will find that they are talking about high-altitude tests, about broken bones, about could sea water be made to be drinkable without removing all the salt? Could people be sterilized quickly and in large numbers? and so on. Now I must explain to you that

*Heinz Norden, trans. (New York: Henry Schuman, 1949).

it wasn't even very good technology. The very fact that it was finally only possible to round up 23 doctors with any claims to distinction shows that not many people of any distinction took part in this. But what's wrong with the whole of that attitude is the notion that the technological expert is allowed to carry out his experiments for the purpose, in particular, of making war, and never mind the cost. By contrast with that, I want to discuss people who try to do something which might be called science in the camps.

People such as Mengele, Entress, Professor Clauberg, worked in the concentration camps particularly on experiments that tried to sterilize people in large numbers. Now it's difficult for me or for any scientist to discuss dispassionately the dreadful stupidity of the work that was done. People like Clauberg had rows of female prisoners brought in, produced some kind of gunk which contained some proportion of acid, squirted it up the women's vaginas, and then asked, "Is it going to work or is it not?" And that was called scientific experimentation.

Now it's not much use talking about whether scientists have a conscience or not when confronted with the absolute scum of the scientific profession that these people represented. They were people who were incapable of devising a single sensible experiment. However, actually an enormous amount of experimentation is going on in this country on the question of how do you, in fact, prevent people from having children. Nearly all the work that's being done at the moment centers on an analysis of the fine function of the hypothalamus, because it's that which controls the ovulatory cycle. And even the pills that are being used at present work not directly but only indirectly through the part of the brain in which the hypothalamus is located. That's the kind of experiment which leads you somewhere. That's not the kind of experiment that the Nazis did.

"All right," you might say, "Perhaps you can't do experiments on human beings. Perhaps a real scientist has a kinship with humanity which doesn't allow him to do experiments on human beings." I'm afraid that's not true. I'm afraid that if you were sufficiently elect and a good scientist, and you really had 10,000 people in your camp who were going to die anyway, and somebody said to you, "do something useful with the prisoners," there are quite a number of important scientific questions still unanswered and that will be unanswered for many years to come toward which we could have made some approach. Let me take a very primitive one. We know that a great many cancers in animals are caused by viruses. We do not yet know of a single example of a virus-induced cancer in human beings. You really don't have to be a very clever scientist to say, "They're all going to be put in the ovens anyway, let's inject half of them with viruses," and so on. I can only tell you that Professor Clauberg and Mr. Mengele had no notion of the reach of proper experimentation. But actually there are much more subtle and profound experiments that we very badly need to do. You see, human beings are very special in that the two sides of the brain carry out very unequal functions; we don't know what they are.

Now and again we may have the good fortune to be able to carry out such experiments. Sometimes there is some piece of brain damage which makes it necessary to separate the two hemispheres. And the beautiful experiments of Roger Sperry at Cal Tech, for instance, or the work of Sir John Eccles give us what little information there is. In general, experiments on brain function are very difficult to carry out; we have practically no information. And certainly we have no information which we can get from anyone but human beings on questions about the bifurcation of the brain. At the moment, we can only get a patient who has had extensive brain damage if, for instance, he's been in a car accident. Then he's in a desperate state where experiments can be done—in the sense that we can actually observe what's going on.

If you really had a concentration camp full of people, and you really were free to do experiments, and you really felt yourself personally free to do them, there are very sensible experiments about the brain you could carry out. That's not what the Nazi's did. The experiments they carried out were experiments that you wouldn't even do on an animal—they're so stupid. They're so stupid because they're wholly removed from the spirit of scientific research. And they're wholly removed from that spirit because they are wholly removed from the sense of human kinship which motivates all this. They start with this notion that there is a magic that we are going to discover.

So I come to say at the end of these chapters why I used the word magic at the beginning—why I want to reach the word civilization at the end. Science is now an integral part of human sensibility. It is in every sense human as an extension of language and as a search for those things which make us specifically ourselves. I hope I've shown that they are very profound things, and I hope I've shown that science in fact cannot be carried on without them. But we do live in a difficult intellectual time—in a time in which it's hard to grasp what's going on and it's terribly easy to leave it to other people to do it, and to say to yourself, "Well, at least, I'm a magician with words" (Yeats), or "I'm a magician with a spinning-jenny."

No kind of magic will do. We have to establish a unitary sense of the human situation, of the fact that cognitive knowledge is the one thing that human beings have been endowed with. That this has made us the only animal that does not fit into an evolutionary niche, but that carves out its own, that makes its physical and mental and cultural environment. And a crucial part of making that cultural environment is to see that the only plan we follow is the great, unbounded, ethical plan of a set of values by which we direct our actions because that's how we are. That is the way to be human.